# Interactive
## Mathematics Program®

INTEGRATED HIGH SCHOOL MATHEMATICS

# High Dive

FIRST EDITION AUTHORS:
Dan Fendel, Diane Resek, Lynne Alper, and Sherry Fraser

CONTRIBUTORS TO THE SECOND EDITION:
Sherry Fraser, IMP for the 21st Century
Jean Klanica, IMP for the 21st Century
Brian Lawler, California State University San Marcos
Eric Robinson, Ithaca College, NY
Lew Romagnano, Metropolitan State College of Denver, CO
Rick Marks, Sonoma State University, CA
Dan Brutlag, Meaningful Mathematics
Alan Olds, Colorado Writing Project
Mike Bryant, Santa Maria High School, CA
Jeri P. Philbrick, Oxnard High School, CA
Lori Green, Lincoln High School, CA
Matt Bremer, Berkeley High School, CA
Margaret DeArmond, Kern High School District, CA

Key Curriculum Press

Second Edition   I M P

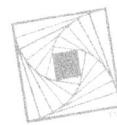

This material is based upon work supported by the National Science Foundation under award numbers ESI-9255262, ESI-0137805, and ESI-0627821. Any opinions, findings, and conclusions or recommendations expressed in this publication are those of the authors and do not necessarily reflect the views of the National Science Foundation.

Key Curriculum Press
1150 65th Street
Emeryville, California 94608
email: editorial@keypress.com
www.keypress.com
10 9 8 7 6 5 4 3 2 1     14 13 12 11
ISBN 978-1-60440-051-9
Printed in the United States
of America

Project Editors
Mali Apple, Josephine Noah, Sharon Taylor

Project Administrators
Emily Reed, Juliana Tringali

Professional Reviewers
Rick Marks, Sonoma State University, CA
D. Michael Bryant, Santa Maria High School, CA, retired

Accuracy Checker
Carrie Gongaware

First Edition Teacher Reviewers
Daniel R. Bennett, Moloka'i High School, HI
Maureen Burkhart, Northridge Academy High School, CA
Dwight Fuller, Ponderosa High School, CA
Daniel S. Johnson, Silver Creek High School, CA
Brian Lawler, California State University San Marcos, CA
Brent McClain, Vernonia School District, OR
Susan Miller, St. Francis of Assisi Parish School, PA
Amy C. Roszak, Cottage Grove High School, OR
Carmen C. Rubino, Silver Creek High School, CA
Barbara Schallau, East Side Union High School District, CA
Kathleen H. Spivack, Wilbur Cross High School, CT
Wendy Tokumine, Farrington High School, HI

First Edition Multicultural Reviewers
Genevieve Lau, Ph.D., Skyline College, CA
Arthur Ramirez, Ph.D., Sonoma State University, CA
Marilyn Strutchens, Ph.D., Auburn University, AL

Copyeditor
Brandy Vickers

Interior Designer
Marilyn Perry

Production Editor
Andrew Jones

Production Director
Christine Osborne

Editorial Production Supervisor
Kristin Ferraioli

Compositor
Lapiz Digital Services, Kristin Ferraioli

Art Editor/Photo Researcher
Maya Melenchuk

Technical Artists
Lapiz Digital Services, Laurel Technical Services, Maya Melenchuk

Illustrators
Taylor Bruce, Deborah Drummond, Tom Fowler, Briana Miller, Evangelia Philippidis, Sara Swan, Diane Varner, Martha Weston, April Goodman Willy

Cover Designer
Jeff Williams

Printer
Lightning Source, Inc.

Mathematics Product Manager
Elizabeth DeCarli

Executive Editor
Josephine Noah

Publisher
Steven Rasmussen

# CONTENTS

## High Dive—Circular Functions and the Physics of Falling Objects

# High Dive

Circular Functions and the Physics of Falling Objects

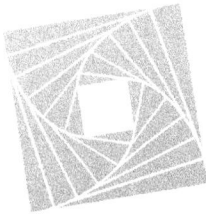

# High Dive—Circular Functions and the Physics of Falling Objects

# Going to the Circus

The central problem of this unit involves a circus act. In the act, a diver falls from a turning Ferris wheel into a tub of water carried by a moving cart. To solve the problem—which involves various kinds of motion—you will need to learn quite a bit of new mathematics.

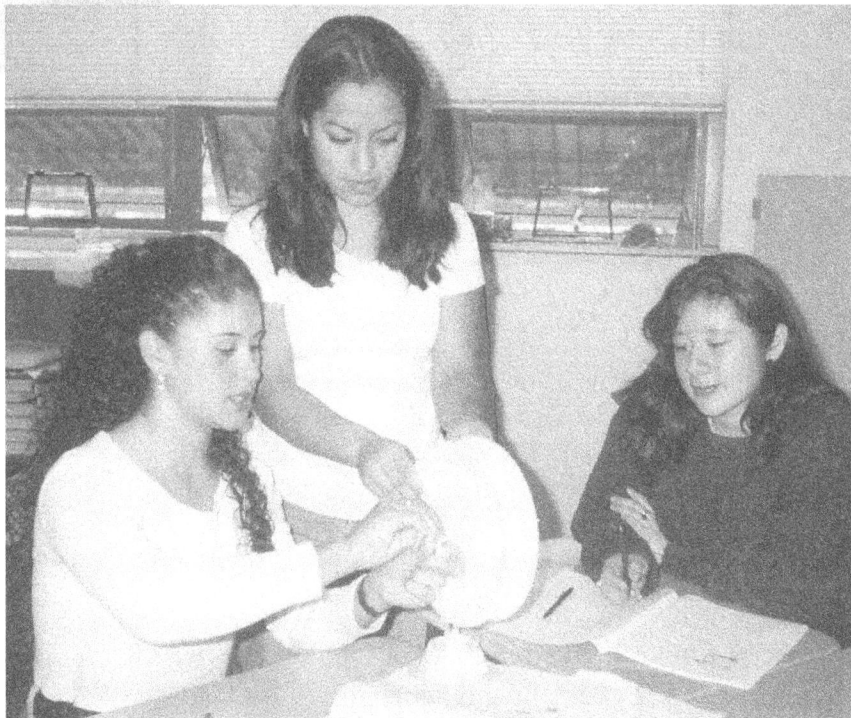

*Maribel DeLoa, Vivian Barajas, and Caroline Moo build a physical model as a first step toward solving the unit problem.*

# The Circus Act

You may have seen or heard about a circus act in which someone dives off a high platform into a small tub of water. The Interactive Circus Troupe has come up with an exciting new variation on this act.

They have attached the diver's platform to one of the pivoting seats on a Ferris wheel so that the platform stays parallel to the ground. The platform sticks out, perpendicular to the plane of the Ferris wheel. The tub of water is on a moving cart that runs along a track, in the plane of the Ferris wheel, and passes under the end of the platform.

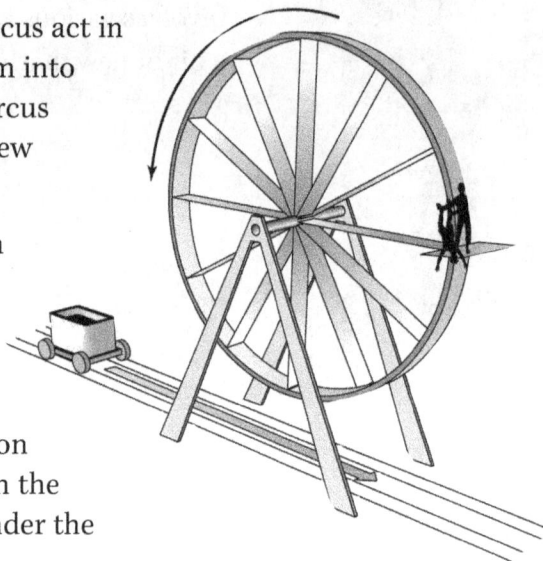

As the Ferris wheel turns, an assistant holds the diver by the ankles. The assistant must let go at exactly the right moment, so that the diver will land in the moving tub of water.

If you were the diver, would you want to trust your assistant's on-the-spot judgment? A slight error and you could go "Splat!" instead of "Splash!"

## Your Task

The diver has insisted that the circus owners hire your group to advise the assistant. You need to figure out exactly when the assistant should let go. Your analysis will be tested on a dummy before it is used with an actual human being.

1. Make a physical model of the problem, using materials that your teacher provides.

2. Specify any other information you need to know about the circus act to determine when the assistant should let go.

*continued* ▶

*Historical note:* The first Ferris wheel was created for the 1893 Chicago World's Fair and was the brainchild of George Washington Gale Ferris. This creation was much larger than most Ferris wheels of today. It stood 265 feet high and was 250 feet in diameter. It carried 36 cars, each of which could hold 60 people. A single revolution took about 20 minutes. Admission was 50¢, ten times the cost of any other ride at the fair. The Ferris wheel was dismantled after the fair and made brief appearances at other major events before being sold for scrap metal in 1906.

# The Tower of Hanoi

## ○ The Legend of the Golden Discs

Buddhism, one of the world's major religions, has roots in India and is practiced by over 300 million people throughout the world. An ancient legend describes an important task once given to a group of Buddhist monks.

According to the legend, a Buddhist temple contained 64 golden discs piled one on top of another. Each successive disc was slightly smaller than the one below it. This pile of discs sat upon a golden tray. Two empty golden trays lay next to this tray.

The monks' task was to move the pile of 64 discs from its original tray to one of the other trays. To do so, they had to follow certain rules. The monks could move only one disc at a time, taking it off the top of a pile and placing it either on an empty tray or on top of an existing pile on one of the trays. Moreover, a disc placed on top of an existing pile could not be larger than the disc below it.

The legend concludes with the promise that when the monks finish this task, the world will be filled with peace and harmony. As you will realize, the monks could not possibly have finished the task. (How unfortunate for the world!)

continued ▶

A famous mathematical puzzle, known as the Tower of Hanoi, is based on this legend. (Hanoi is the capital of Vietnam, which is in Southeast Asia. Many people in Vietnam are Buddhists.) The puzzle consists of three pegs and a set of discs of different sizes, as shown in the diagram.

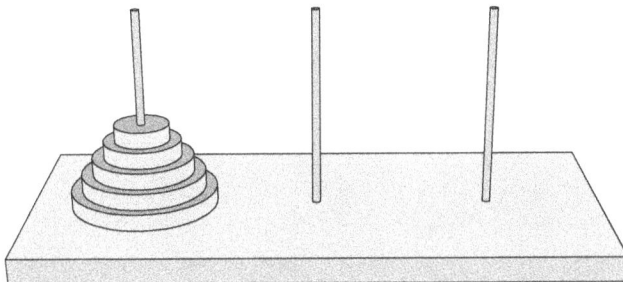

The discs have holes in their centers. To begin with, the discs are all placed over the peg at the left, with the largest disc on the bottom and with the discs decreasing in size as they go up. (This diagram shows only 5 discs instead of 64.)

The task in this puzzle is to transfer all 5 of the discs to the peg on the right. As in the legend, the discs must be moved according to certain rules.

• Only one disc can be moved at a time.

• The disc being moved must be the top disc on its peg.

• The disc being moved must be placed either on an empty peg or on top of a larger disc on a different peg.

## ○ Getting Started

Your POW is to answer this question.

*If the monks move one disc every second, how long will it take them to complete their task?*

Start your investigation of this question with just two or three discs. Work your way up, finding the *least number* of moves required to transfer the pile of discs from the peg on the left to the peg on the right.

*continued* ▶

As you work, consider these questions. The notation $a_n$ represents the number of moves required to move $n$ discs from the peg on the left to the peg on the right.

- If you know how many moves are needed to move 20 discs, how can you find the number of moves needed to move 21 discs? Can you generalize this process into a formula? That is, if you know $a_n$, how can you find $a_{n+1}$? Can you explain why this formula holds true?

- Look for a formula that gives $a_n$ directly in terms of $n$. Test your formula with specific cases. If you knew that this formula worked for $n = 20$, could you prove that it worked for $n = 21$? Can you prove the formula in general?

When you have answered these questions as best you can, return to the question about the monks and their 64 discs.

○ *Write-up*

1. *Process*

2. *Results:* Give the results of your investigation, including these details.

   - The number of moves required for any specific cases you studied
   - The amount of time required for the monks to move the 64 discs
   - Any general formulas or procedures that you found, even if you aren't sure of them

3. *Solutions:* Explain your results, including how you know that the number of moves for each number of discs is the least possible. Also give any explanations you found for your generalizations.

4. *Self-assessment*

# The Ferris Wheel

Al and Betty are at the amusement park to ride on a Ferris wheel. This wheel has a radius of 15 feet, and its center is 20 feet above ground level.

You can describe various positions in the cycle of a Ferris wheel in terms of the face of a clock, as indicated in the diagram. For example, the highest point in the wheel's cycle is the 12 o'clock position, and the point farthest to the right is the 3 o'clock position.

For simplicity, think of Al and Betty's location as they ride as simply a point on the circumference of the wheel's circular path. That is, ignore the size of the Ferris wheel seats, the heights of Al and Betty, and so on.

1. How far off the ground are Al and Betty when they are at each of these positions?

   a. The 3 o'clock position

   b. The 12 o'clock position

   c. The 9 o'clock position

   d. The 6 o'clock position

2. How far off the ground are Al and Betty when they are at the 2 o'clock position? (*Caution:* Their height at the 2 o'clock position is not a third of the way between their height at the 3 o'clock position and their height at the 12 o'clock position.)

3. Pick two other clock positions. Figure out how far off the ground Al and Betty are when they reach each of those positions.

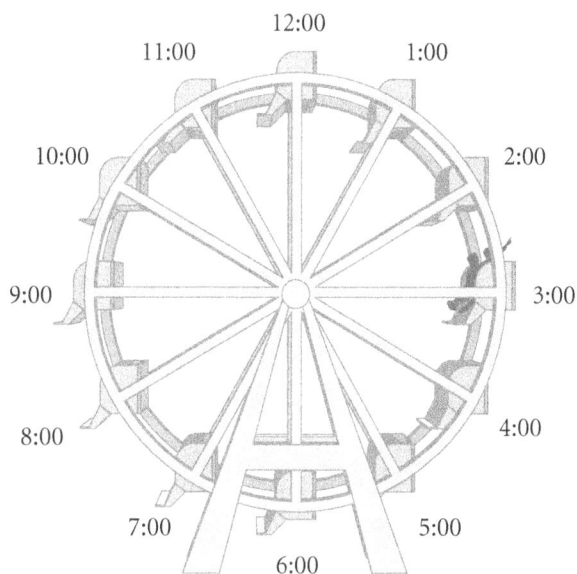

# As the Ferris Wheel Turns

To understand what happens when a diver is released from a moving Ferris wheel, you need precise information about the position of the diving platform as the Ferris wheel turns.

In this activity, you will look only at the *height* of the platform. Later, you will consider how far the platform is to the left or right of the center of the wheel.

You will need this information about the Ferris wheel.

- The radius of the Ferris wheel is 50 feet.

- The wheel turns at a constant speed, makes a complete turn every 40 seconds, and moves counterclockwise.

- The center of the wheel is 65 feet off the ground.

Use these facts throughout the unit, unless a problem specifically gives different information. *Reminder:* The circumference of a circle can be found from its radius using the formula $C = 2\pi r$.

1. At what speed is the platform moving (in feet per second) as it goes around on the Ferris wheel?

2. The rate at which an object turns is called *angular speed,* because it measures how fast an angle is changing. Angular speed does not depend on the radius. Through what angle (in degrees) does the Ferris wheel turn each second?

*continued* ▶

3. How many seconds does it take for the platform to go each of these distances?

   a. From the 3 o'clock to the 11 o'clock position

   b. From the 3 o'clock to the 7 o'clock position

   c. From the 3 o'clock to the 4 o'clock position

4. What is the platform's height off the ground at each of these times?

   a. 1 second after passing the 3 o'clock position

   b. 6 seconds after passing the 3 o'clock position

   c. 10 seconds after passing the 3 o'clock position

   d. 14 seconds after passing the 3 o'clock position

   e. 23 seconds after passing the 3 o'clock position

   f. 49 seconds after passing the 3 o'clock position

## The Height and the Sine

You've seen that trigonometric functions can be helpful in describing where the platform is as it travels around on the Ferris wheel. But the basic right-triangle definitions of these functions work only for acute angles.

In the next several activities, you'll explore how to extend the definition of the sine function to arbitrary angles. You will also investigate how to use this extended definition to get a general formula for the platform's height.

*Jason Weinstock presents the formula his group developed to express Al and Betty's height off the ground as the Ferris wheel turns.*

# At Certain Points in Time

In *As the Ferris Wheel Turns*, you found the height of the platform after it had turned for specific amounts of time. You probably saw that this is easiest when the platform is in the first quadrant.

Now you will generalize your work for the case of the first quadrant. The basic facts about the Ferris wheel are the same as in *As the Ferris Wheel Turns*. In particular, the period is 40 seconds, so the platform remains in the first quadrant for the first 10 seconds.

1. Suppose the Ferris wheel has been turning for $t$ seconds, with $0 < t < 10$. Represent the platform's height off the ground as $h$, and find a formula for $h$ in terms of $t$.

2. Verify your formula, which is for the first quadrant, using your results from Questions 4a ($t = 1$) and 4b ($t = 6$) of *As the Ferris Wheel Turns*.

# A Clear View

As you may remember, the Ferris wheel at the amusement park where Al and Betty like to ride has a radius of 15 feet, and its center is 20 feet above ground level. This is not the same Ferris wheel as the one at the circus.

The park's Ferris wheel turns with a constant angular speed. It takes 24 seconds for a complete turn.

The fence around the amusement park is 13 feet high. Once Al and Betty get above the fence, there is a wonderful view.

1. During one revolution, what percentage of the time are Al and Betty above the height of the fence?

2. How would your answer to Question 1 change if the period were other than 24 seconds?

If the Ferris wheel at the circus turns counterclockwise at a constant angular speed of 9 degrees per second and the platform passes the 3 o'clock position at $t = 0$, then the platform will remain in the first quadrant through $t = 10$.

During this time interval, the platform's height above the ground is given by the formula

$$h = 65 + 50 \sin (9t)$$

However, the right-triangle definition of the sine function makes sense only for acute angles. To make this formula work for all values of $t$, we need to extend the definition of the sine function to include all angles.

## The Coordinate Setting

The context of the Ferris wheel could be used to develop this extended definition. However, the standard approach uses a more abstract setting that makes it easier to apply the definition to other situations.

The angle $\theta$ is placed within a coordinate system, with its vertex at the origin. The angle is measured counterclockwise from the positive direction of the $x$-axis. The goal is to express $\sin \theta$ in terms of the $x$- and $y$-coordinates of a point on the ray defining the angle, such as point $A$ in the first diagram. $A$ is assumed to be different from the origin.

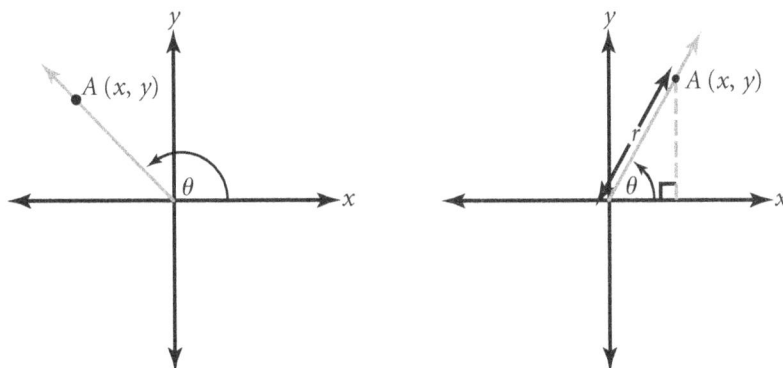

*continued*

When $\theta$ is an acute angle, we get a diagram like the second one. It's helpful to introduce the letter $r$ to represent the distance from $A$ to the origin. This distance is also the length of the hypotenuse of the right triangle.

Based on the right-triangle definition of the sine function, we get

$$\sin \theta = \frac{y}{r}$$

Mathematicians use this equation to extend the definition of the sine function to arbitrary angles. That is, they define $\sin \theta$ as the ratio $\frac{y}{r}$ for *any* angle $\theta$. (This automatically means that the new definition agrees with the old one for acute angles.)

## The Ferris Wheel Analogy

You can think of $A$ as a point on the circular path of the Ferris wheel, as shown in this diagram. In this context, $r$ corresponds to the radius of the Ferris wheel, and $y$ corresponds to the platform's height *relative to the center of the Ferris wheel.*

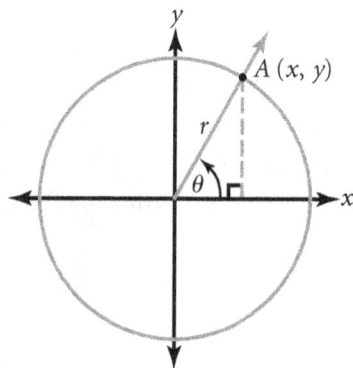

# Testing the Definition

You've seen that the sine function can be extended to all angles using the *xy*-coordinate system. The big question is this.

*If you use this coordinate definition of the sine function, does the platform-height formula work for all angles?*

In this activity, you will investigate that question.

1. If the platform has been turning for 25 seconds, it has moved through an angle of 225° and is in the third quadrant of its cycle.

   a. Use a diagram like the one shown here to find the value of sin 225° based on the coordinate definition of the sine function. First choose a specific value for *r*. Then find the value of *y* using the right triangle.

   b. Substitute your answer from part a into the expression 65 + 50 sin 225°.

   c. Explain why your answer to part b is reasonable for the position of the platform after 25 seconds.

   d. Verify that your calculator gives the same value for sin 225° that you found in part a.

2. Go through a sequence of steps like those in Question 1 using the value *t* = 32, which places the platform in the fourth quadrant. You will first need to find the actual height of the platform for *t* = 32.

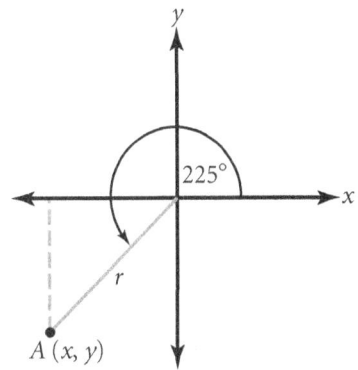

# Graphing the Ferris Wheel

1. Plot individual points to create a graph showing the platform's height, $h$, as a function of the time elapsed, $t$. Explain how you get the value for $h$ for each point you plot. Your graph should show the first 80 seconds of the Ferris wheel's movement. Use the basic information about the Ferris wheel from *As the Ferris Wheel Turns*.

2. Describe how this graph would change if you made each of the adjustments described in parts a to c of this question. Treat each part separately, changing only the item mentioned and keeping the rest of the information as in Question 1.

   a. How would the graph change if the radius of the Ferris wheel were smaller?

   b. How would the graph change if the Ferris wheel were turning faster—that is, if the period were shorter?

   c. How would the graph change if you measured height with respect to the center of the Ferris wheel instead of with respect to the ground? For example, if the platform were 40 feet above the ground, you would treat this as a height of $-25$ feet, because 40 feet above the ground is 25 feet below the center of the wheel.

# Ferris Wheel Graph Variations

In Question 1 of *Graphing the Ferris Wheel,* you made a graph showing how the height of the platform depends on the time that has elapsed since the wheel began moving. That graph was based on the "standard" Ferris wheel, which has a radius of 50 feet, a period of 40 seconds, and a center that is 65 feet off the ground. The diagram shows two periods of that graph.

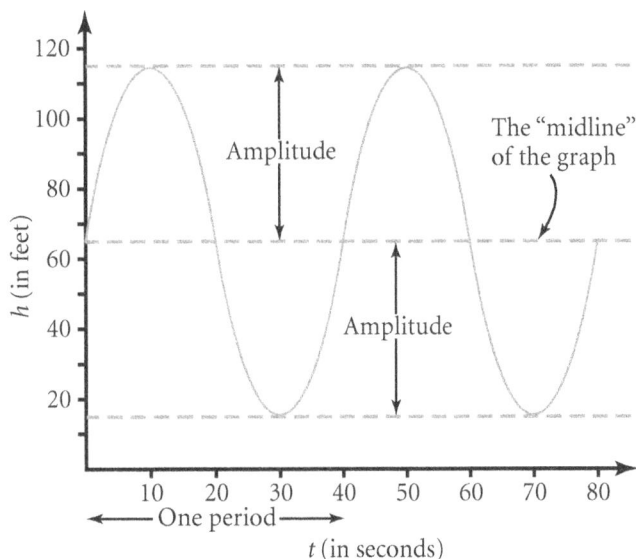

The dashed line at $h = 65$ shows the "midline" of the graph. The graph is as much above this line as it is below. The other dashed lines, at $h = 115$ and $h = 15$, show the maximum and minimum heights of the graph. The distance from the midline to the maximum or minimum is called the **amplitude** of the graph. The amplitude of this graph is 50.

In Question 2 of *Graphing the Ferris Wheel,* you described how the graph would change if you made certain adjustments to the Ferris wheel. In this activity, you will look at those changes in more detail.

Treat each question as a separate problem, changing only the item mentioned and keeping the rest of the information as in the standard Ferris wheel. Use the same scale for all your graphs, to make visual comparisons easier.

1. a. Pick a new value, less than 50 feet, for the radius, and draw the graph.

   b. Give an equation for your new graph, expressing $h$ (the height of the platform, in feet) in terms of $t$ (the time elapsed, in seconds).

   c. Pick a value for $t$, and verify that your equation from part b gives the value you used in your graph for that value of $t$.

*continued* ▶

2. a. Pick a new value, less than 40 seconds, for the period, and draw the graph.

   b. Give an equation for your new graph, expressing $h$ in terms of $t$.

   c. Pick a value for $t$, and verify that your equation from part b gives the value you used in your graph for that value of $t$.

3. Suppose the Ferris wheel is set up inside a large hole so that its center is exactly level with the ground.

   a. Draw the graph based on this change.

   b. Give an equation for your new graph, expressing $h$ in terms of $t$.

   c. Pick a specific value for $t$, and verify that your equation from part b gives the value you used in your graph for that value of $t$.

# The "Plain" Sine Graph

The height of the Ferris wheel platform is given by a formula that involves the sine function. In previous activities, you've graphed this height function and examined how the graph changes as details of the Ferris wheel are changed.

Now you'll look at the graph of the "plain" sine function—outside of the context of the Ferris wheel.

1. Draw the graph of the function defined by the equation $z = \sin \theta$ for values of $\theta$ from $-360°$ to $720°$. (*Note:* To avoid confusion with $x$- and $y$-coordinates or the idea that $t$ represents time and $h$ represents height on the Ferris wheel, we are introducing new variables here.)

2. What is the amplitude of this function?

3. What is the period of this function? Why is the sine function periodic?

4. What are the $\theta$-intercepts of the graph?

5. What values of $\theta$ make $\sin \theta$ a maximum? What values of $\theta$ make $\sin \theta$ a minimum?

6. Suppose the equation $h = \sin t$ describes the platform-height function for some Ferris wheel. What are the specifications of that Ferris wheel? That is, what are its radius, its period, and the height of its center? Indicate any ways in which this wheel differs from the standard Ferris wheel described in *As the Ferris Wheel Turns*.

# Sand Castles

Oceana loves to build elaborate sand castles. Her problem is that her castles take a long time to build and often get swept away by the incoming tide.

Oceana is planning a trip to the beach next week. She decides to pay attention to the tides so that she can plan her castle building and have as much time as possible.

The beach slopes up gradually from the ocean toward the parking lot. Oceana considers the waterline to be "high" if the water comes farther up the beach, leaving less sandy area visible. She considers the waterline to be "low" if there is more sandy area visible. Oceana likes to position herself as close to the water as possible because damp sand is better for building.

According to Oceana's analysis, the water level on the beach for the day of her trip will fit this equation.

$$w(t) = 20 \sin (29t)$$

In this equation, $w(t)$ represents how far the waterline is above or below its average position. The distance is measured in feet, and $t$ represents the number of hours elapsed since midnight.

*continued* ▶

In the case shown in the diagram, the waterline is above its average position, and $w(t)$ is positive.

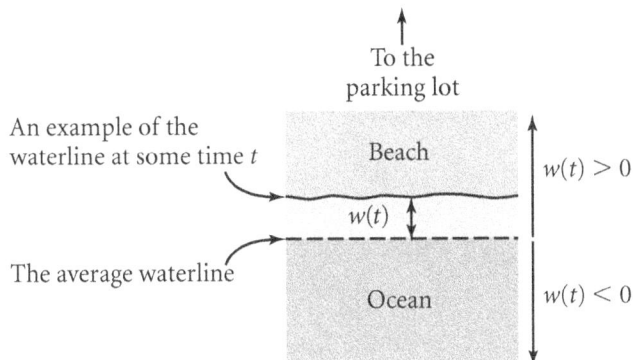

1. Graph the waterline function for a 24-hour period.

2. a. What is the highest position up the beach (compared to its average position) that the waterline will be during the day? (This is called *high tide*.)

   b. What is the lowest position that the waterline will be during the day? (This is called *low tide*.)

3. Suppose Oceana plans to build her castle right on the average waterline just as the water has moved below that line. How much time will she have to build before the water returns and destroys her work?

4. Suppose Oceana wants to build 10 feet below the average waterline. What is the maximum amount of time she can have for making her castle?

5. Suppose Oceana decides she needs only two hours to build and admire her castle. What is the lowest position on the beach where she can build?

# Paving Patterns

Al and Betty are helping Al's family lay paving stones for
a path along the side of their house. The path will be exactly
2 feet wide. Each paving stone is rectangular, with dimensions
1 foot by 2 feet.

You might think this would be easy: simply lay one stone after
another across the path. But there is more than one way to place
the stones.

For example, a section of the path 3 feet long could use any of these
three arrangements. *Important:* These arrangements are all considered
different, even though the first two are very much alike.

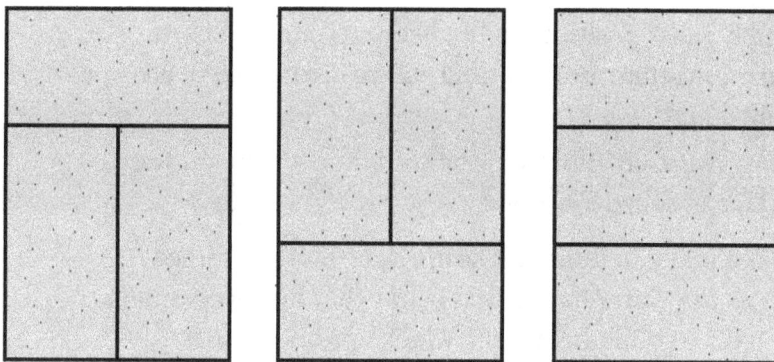

Al and Betty want to know how many different ways there are to lay
out the stones. The path is 20 feet long.

Al and Betty start to analyze the situation by using 1-inch-by-2-inch
plastic tiles set inside a 2-inch-by-20-inch rectangle, but they are soon
overwhelmed by all the possibilities.

Can you help? You might want to start with shorter paths and look
for patterns in the number of cases.

You do not need to show the patterns themselves, except to
explain your thinking. Instead, focus on *how many* patterns
there are for a path of a given length.

*continued*

○ *Write-up*

1. *Problem statement*

2. *Process*

3. *Results*

   - Give the numeric results for any specific cases you studied.
   - Give any general formulas you found, even if you aren't sure of them.
   - Give any explanations you found for your formulas.

4. *Self-assessment*

# More Beach Adventures

1. After spending the day building sand castles, Oceana wants to take an evening walk with a friend along the shoreline.

   Oceana knows that one stretch along the shore is quite rocky. At that point, the rocks jut into the ocean. To walk around them, a person has to follow a path that is 14 feet below the average waterline.

   If Oceana and her friend don't want to get wet, they need to take their walk when the waterline is 14 feet or more below the average waterline. What is the time period during which they can take their walk?

   Remember that the position of the waterline over the course of the day is given by the equation $w(t) = 20 \sin (29t)$, where the distance is measured in feet and $t$ represents the number of hours elapsed since midnight.

2. While she builds sand castles, Oceana likes to amuse herself by looking for numbers that have a sine of a given value. Here are four problems she thought about recently. In parts a and b, find exact values for $\theta$. In parts c and d, give $\theta$ to the nearest degree. Your solutions should be between $-360°$ and $360°$.

   a. Find three values of $\theta$, other than 15°, such that $\sin \theta = \sin 15°$.

   b. Find three values of $\theta$ such that $\sin \theta = -\sin 60°$.

   c. Find three values of $\theta$ such that $\sin \theta = 0.5$.

   d. Find three values of $\theta$ such that $\sin \theta = -0.71$.

# Falling, Falling, Falling

The diver on the Ferris wheel doesn't simply go round and round. At some point, the assistant lets go and the diver begins his fall.

How long will the diver be in the air? You'll have to learn some principles of physics to answer this question. The question is complicated by the fact that the diver does not fall at a constant speed.

*Stephanie Lin records a formula for the height of an object falling from rest.*

# Distance with Changing Speed

1.  Curt is traveling home from college to visit his family. He drives from 1 p.m. to 3 p.m. at an average speed of 50 miles per hour. Then he drives from 3 p.m. to 6 p.m. at an average speed of 60 miles per hour.

    a.  Draw a graph showing Curt's speed as a function of time for the entire period from 1 p.m. to 6 p.m. Treat his speed as constant for each of the two time periods—from 1 p.m. to 3 p.m. and from 3 p.m. to 6 p.m.

    b.  Describe how to use areas in your graph to represent the total distance Curt travels.

2.  A triathlete is running at a steady speed of 20 feet per second. At exactly noon, she starts to increase her speed. Her speed increases at a constant rate so that 20 seconds later, she is going 30 feet per second.

    a.  Graph the runner's speed as a function of time for this 20-second interval.

    b.  Calculate the runner's average speed for this 20-second interval.

    c.  Explain how to use area to find the total distance she runs during this 20-second interval.

# Acceleration Variations and a Sine Summary

## Part I: Acceleration Variations

In Question 2 of *Distance with Changing Speed,* you considered the case of a person running with constant acceleration. In other words, the runner's speed was increasing at a constant rate.

In that question, the runner's speed went from 20 feet per second to 30 feet per second over a 20-second time interval. The graph shows the runner's speed as a function of time.

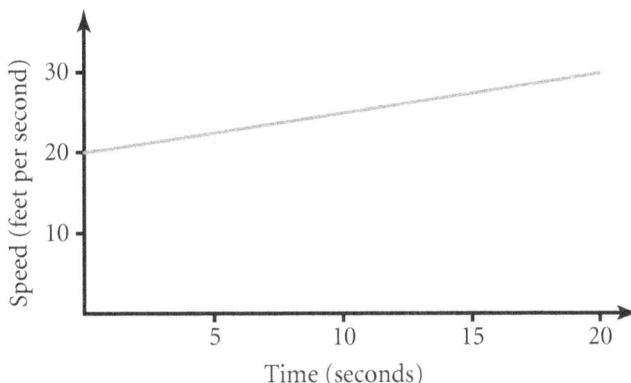

That question illustrates an important principle.

> If an object is traveling with constant acceleration, then its average speed over any time interval is the average of its beginning speed and its final speed during that time interval.

According to this principle, the runner's average speed for the 20-second interval was exactly 25 feet per second, which is the average of 20 feet per second (the beginning speed) and 30 feet per second (the final speed).

Your first task in this activity is to describe three variations on this situation. In each case, the runner's speed should increase, as before, from 20 feet per second to 30 feet per second over the same 20-second time interval. But in your examples, *the runner's acceleration should not be constant.*

*continued*

For each example, graph the runner's speed in terms of time.

- Give an example in which the runner's average speed is *more than* the average of her beginning speed and her final speed.

- Give an example in which the runner's average speed is *less than* the average of her beginning speed and her final speed.

- Give an example in which the runner's average speed is *equal to* the average of her beginning speed and her final speed. Remember that even with this example, her acceleration should not be constant.

## Part II: A Sine Summary

The idea of extending the sine function to all angles—not merely acute angles—is an important concept. Your task now is to reflect on your work with the sine function. Include these things in your written work.

- A summary of what you have learned so far about this idea

- Any questions you still have about this extended sine function

- An explanation of how the extension of the sine function helps with the solution of the unit problem

# Free Fall

Since the beginning of time, objects have fallen. But it wasn't until the sixteenth and seventeenth centuries that scientists understood the physics and mathematics of falling objects.

The Italian physicist Galileo Galilei (1564–1642) is one of those credited with figuring out the laws of gravitational fall based on experiments. The English physicist Isaac Newton (1642–1727) developed a broader theory of gravitation to explain Galileo's observations.

## Free-Falling Objects

Using experiments and theoretical analysis, physicists have confirmed this principle.

Falling objects have constant acceleration.

This principle assumes there is no air resistance or other complicating factors to interfere with an object's fall. That is, the principle describes the behavior of *free-falling* objects. In this unit, assume that unless you are told otherwise, falling objects are falling freely.

This broad principle of free-falling objects can be stated more precisely.

The instantaneous speed of a freely falling object increases approximately 32 feet per second for each second of the object's fall.

## Starting from Rest

The simplest case of a free-falling object is when an object starts from rest—that is, when its speed is zero at $t = 0$. In this case, the object's instantaneous speed after 1 second is 32 feet per second; after 2 seconds, its instantaneous speed is 64 feet per second; and so on.

*continued*

## From Acceleration to Distance

Your task is to use the principles just stated to express the distance an object falls in terms of the amount of time it has been falling. Assume the object is dropped from rest and falls freely.

1. a. How fast is the object going at $t = 5$?

   b. How far does the object fall in the first 5 seconds?

2. Generalize your work from Question 1 to develop a formula for how far the object falls in the first $t$ seconds.

3. Suppose the object starts falling from a height of $h$ feet. What is its height after $t$ seconds? Assume the object has not yet reached the ground.

4. Use your result from Question 3 to find an expression, in terms of $h$, for the amount of time it would take for the object to reach the ground.

Now apply your work to a simple version of the circus act.

5. Suppose the platform is fixed at 90 feet above the ground, the diver falls freely from rest, and the water level in the tub is 8 feet above the ground. How long will it take the diver to reach the water?

# Not So Spectacular

The circus owner decides that to save money, he will fill in for the Ferris wheel diver from time to time.

This isn't actually a very good idea. The owner is not an experienced diver, so he can't safely fall great distances. In fact, he refuses to be dropped from more than 25 feet above the ground. He also insists there be a huge tub of water under him at all times.

Your task is to find all possible times when the owner will be 25 feet from the ground. You may want to describe the complete set of possibilities by writing an algebraic expression for $t$.

Here are the basic facts about the Ferris wheel.

- The radius is 50 feet.
- The center of the wheel is 65 feet off the ground.
- The wheel turns counterclockwise at a constant angular speed, with a period of 40 seconds.
- The platform is at the 3 o'clock position when the Ferris wheel starts moving.

# A Practice Jump

After some not-so-high practice dives by the circus owner, the circus performers decide to do a practice run of the show with the diver himself. But they decide to set it up so that they will not have to worry about a moving cart.

Instead, the cart containing the tub of water is placed directly under the Ferris wheel's 11 o'clock position. As usual, the platform passes the 3 o'clock position at $t = 0$.

1. How many seconds will it take for the platform to reach the 11 o'clock position?

2. What is the diver's height off the ground when he is at the 11 o'clock position?

One purpose of this practice run is to find how long it will take for the diver to fall into the water. You should be able to predict this, based on the formula that an object falling freely from rest takes $\sqrt{\frac{h}{16}}$ seconds to fall $h$ feet. Assume the diver is falling freely from rest.

3. How long will it take from the time the diver is released until he hits the water? Don't forget that the water level in the cart is 8 feet above the ground.

4. More generally, suppose the assistant lets go $W$ seconds after the Ferris wheel starts turning. Here, $W$ stands for "wheel time." Assuming the cart is in the right place, how long will the diver be in the air before he hits the water?

# Moving Left and Right

Up to this point, you have mostly been considering the platform's position and the diver's motion in the vertical dimension. But as the Ferris wheel turns, the platform is also moving to the left or right, and the cart is moving steadily to the right (once it gets started).

The key to a successful dive is having the cart at the right place at the right time. It's now time for you to consider the horizontal dimension of the Ferris wheel problem.

*Kevin Brandt and Jordyn Vincelet discuss the fact that the turning Ferris wheel also involves horizontal movement of the diver's platform.*

# Cart Travel Time

So far in the unit, you've focused mainly on the position and motion of the diver. Where's the cart of water in all this?

The cart starts moving when the Ferris wheel passes the 3 o'clock position. The goal is for the cart to be in the correct position when the diver reaches the level of the water in the cart. In this activity, you will consider only the cart's *travel time.*

Suppose the assistant lets go of the diver $W$ seconds after the Ferris wheel passes the 3 o'clock position. Write an expression in terms of $W$ for the length of time the cart will have traveled from the moment it starts until the moment the diver reaches the level of the water.

# Where Does He Land?

Earlier in the unit, you found that the expression $65 + 50 \sin(9t)$ gives the diver's height off the ground while he is still on the platform. But what about the diver's *horizontal* position? This will be crucial in determining whether he lands in the tub of water on the moving cart.

To describe the diver's horizontal position, you will use a horizontal coordinate system, as shown here. In this coordinate system, an object's $x$-coordinate is based on its distance (in feet) to the right or left of the center of the Ferris wheel, with objects to the right of the center having positive $x$-coordinates.

For instance, the platform and diver have an $x$-coordinate of 50 when the platform passes the 3 o'clock position. The cart starts its motion with an $x$-coordinate of $-240$, because it is initially 240 feet to the left of the center of the Ferris wheel's base.

As usual, make these assumptions: The platform passes the 3 o'clock position at $t = 0$. The wheel turns counterclockwise at a constant rate, with a period of 40 seconds. The diver falls straight down once he is released.

1. Where will the diver land if he is released at each of these times?

   a. $t = 3$

   b. $t = 7$

   c. $t = 12$

   d. $t = 26$

   e. $t = 37$

*continued* ▶

2. Sketch a graph of the platform's *x*-coordinate as a function of *t*. Your graph should cover two complete turns of the Ferris wheel— from $t = 0$ to $t = 80$.

*Note:* Although the platform's *x*-coordinate represents its horizontal position, in this context *x* is a function of *t*. That is, *t* is the independent variable and *x* is the dependent variable. That means you should show the value of *t* on the horizontal axis of your graph and the value of *x* on the vertical axis.

# First Quadrant Platform

In Question 1 of *Where Does He Land?* you found the *x*-coordinate of the diver's landing position for five specific cases.

At the moment the diver is released, his *x*-coordinate is the same as the *x*-coordinate of the platform. In Question 2 of *Where Does He Land?* you sketched a graph of the platform's *x*-coordinate as a function of *t*.

Now focus on values of *t* between 0 and 10, so that the platform is still in the first quadrant. Develop an equation that gives the platform's *x*-coordinate in terms of *t*.

# Carts and Periodic Problems

## Part I: Where's the Cart?

In *Where Does He Land?* you looked at the horizontal coordinate of the diver when he falls. You also need to know where the cart is while the diver is falling—and especially where it is when the diver reaches the water.

The cart begins 240 feet to the left of the Ferris wheel's base, so its $x$-coordinate at $t = 0$ is $-240$. The cart moves to the right at 15 feet per second and begins moving at that speed at $t = 0$.

Based on this information, find the cart's $x$-coordinate at the moment the diver reaches the water level.

## Part II: Periodic Problems

You have seen that the height of a platform on a Ferris wheel represents a periodic function. You have encountered periodic functions before. For instance, the swinging of a pendulum is periodic motion, and the bob's distance from the center line is a periodic function of time (assuming the pendulum isn't slowing down).

1. Describe three other situations that you believe are periodic. For each example, explain what is repeating and give the period for the repetition.

2. Sketch graphs of at least two of the periodic situations you described.

# Generalizing the Platform

If the Ferris wheel platform starts at the 3 o'clock position, with the Ferris wheel turning counterclockwise at a constant angular speed of 9 degrees per second, then the platform will remain in the first quadrant through $t = 10$.

During this time interval, the platform's $x$-coordinate is given by this formula.

$$x = 50 \cos (9t)$$

This formula uses the facts that the Ferris wheel's radius is 50 feet and the angular speed is 9 degrees per second. But the right-triangle definition of the cosine function applies only to acute angles, so this formula isn't defined if $t$ is greater than 10. In this activity, you will explore how to extend the definition of the cosine function.

## Specific Cases

1. Consider the case $t = 12$.

   a. Find the platform's $x$-coordinate when $t = 12$. This was Question 1c of *Where Does He Land?* You may want to express your answer in terms of the cosine of some acute angle.

   b. What value should you assign to $\cos (9 \cdot 12)$ so that the formula $x = 50 \cos (9t)$ gives your answer from part a when you substitute 12 for $t$?

*continued*

2. Consider the case $t = 26$.

   a. Find the platform's $x$-coordinate when $t = 26$. This was Question 1d of *Where Does He Land?* You may want to express your answer in terms of the cosine of some acute angle.

   b. What value should you assign to $\cos(9 \cdot 26)$ so that the formula $x = 50 \cos(9t)$ gives your answer from part a when you substitute 26 for $t$?

## The General Case

3. How can you define $\cos \theta$ in a way that makes sense for all angles and that gives the results you needed in Questions 1b and 2b? You may want to look back at the activity *Extending the Sine*.

# Planning for Formulas

You now have all the parts to the puzzle. You simply have to put them together! Suppose $W$ represents the amount of time the Ferris wheel has been turning at the moment the diver is released. You have formulas that tell you each of these things in terms of $W$.

- The diver's height at the moment he is released
- The diver's $x$-coordinate at the moment he is released
- The length of time the diver falls until he reaches the water level
- The cart's $x$-coordinate when the diver reaches the water level

Write out each of these four formulas. Explain each formula clearly, including how the general definitions of sine and cosine and the principles of falling objects are used in them.

Also discuss how each of the following facts fits into your formulas.

- The Ferris wheel has a radius of 50 feet.
- The center of the Ferris wheel is 65 feet above the ground.
- The Ferris wheel turns counterclockwise at a constant rate, making a complete turn every 40 seconds.
- When the cart starts moving, it is 240 feet to the left of the Ferris wheel's base.
- The cart moves to the right along the track at a constant speed of 15 feet per second.
- The water level in the cart is 8 feet above the ground.
- When the cart starts moving, the platform is at the 3 o'clock position.

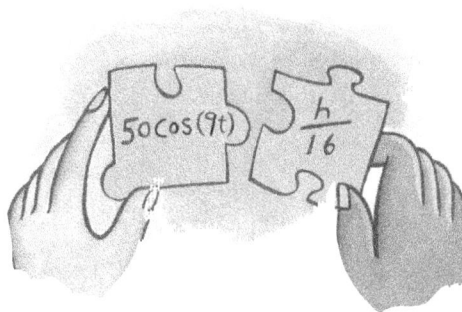

# Finding the Release Time

You have developed a large collection of formulas that explain specific parts of the Ferris wheel problem. Now it's time to put them all together.

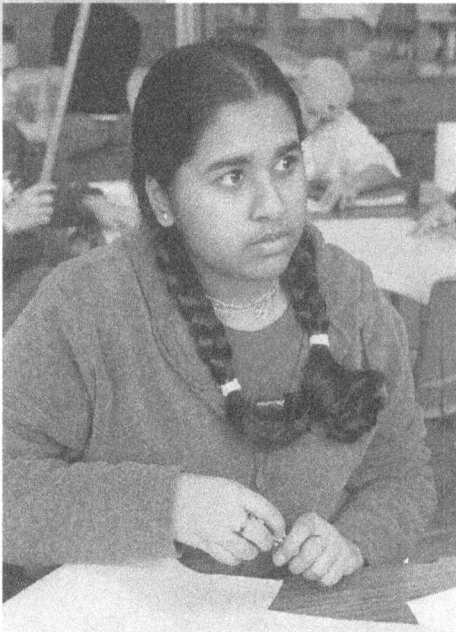

*Shaleen Nand ponders how to pull together all the information she's developed and the various formulas she's collected.*

# Moving Cart, Turning Ferris Wheel

Your task is to figure out when the assistant should let go of the diver. Let $t = 0$ represent the moment when the platform passes the 3 o'clock position. Let $W$ represent the number of seconds until the diver is released. You need to determine the right value for $W$.

In addition to giving the value of $W$, also figure out these things.

* Where the platform will be in the Ferris wheel's cycle when the diver is dropped
* Where the cart will be when the diver hits the water

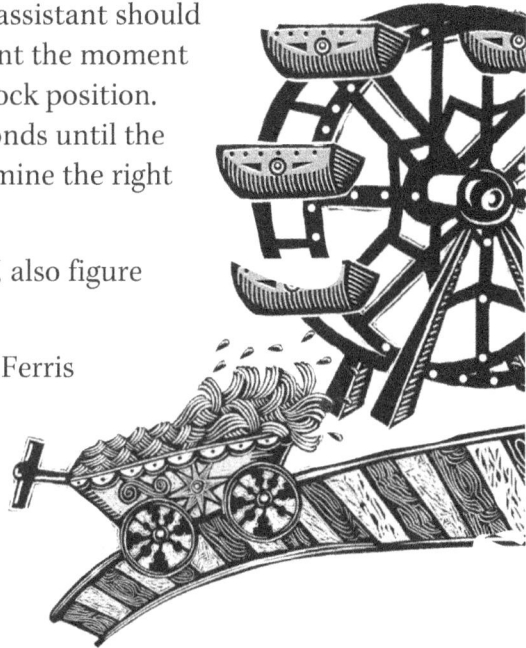

# Putting the Cart Before the Ferris Wheel

What if you could change where the cart started? That might make things a little easier.

In this activity, assume that all the facts about the Ferris wheel and the cart are the same as usual except for the cart's initial position.

Suppose the diver is released exactly 25 seconds after the Ferris wheel begins turning from its 3 o'clock position.

1. What is the diver's *x*-coordinate as he falls?

2. Where should the cart start out so that the diver will fall into the tub of water? Assume the cart still starts to the left of the Ferris wheel and travels to the right at 15 feet per second.

# What's Your Cosine?

You have seen that the cosine function is
defined in a manner similar to that for the
sine function. If $\theta$ is any angle, draw a ray
from the origin, making a counterclockwise
angle of that size with the positive $x$-axis.
Pick a point $(x, y)$ on the ray (other than the
origin). Define $r$ as the distance from $(x, y)$ to
the origin, so $r = \sqrt{x^2 + y^2}$.

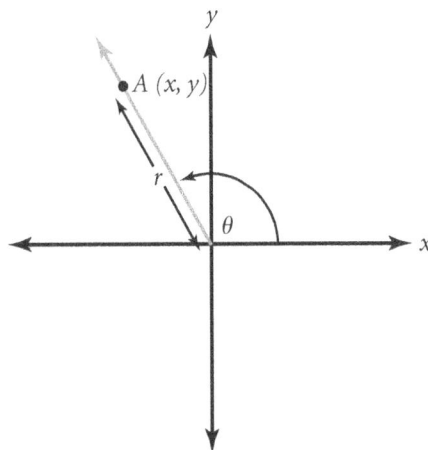

Then define the cosine function for all angles
by this equation.

$$\cos \theta = \frac{x}{r}$$

As with the sine function, this definition gives the same values for
acute angles as the right-triangle definition. In addition, the extended
cosine function can give the same value for different angles.

1. Draw the graph of the function defined by the equation $z = \cos \theta$
   for values of $\theta$ from $-360°$ to $720°$.

   a. What is the amplitude of this function?

   b. What is the period of this function? Why is the cosine function
      periodic?

   c. What are the $\theta$-intercepts of the graph?

   d. What values of $\theta$ make $\cos \theta$ a maximum? What values of $\theta$
      make $\cos \theta$ a minimum?

*continued* ▶

2. These questions are similar to the questions about the sine function in *More Beach Adventures*. As in that activity, your solutions should be between $-360°$ and $360°$. In parts a and b, find exact values for $\theta$. In parts c and d, give $\theta$ to the nearest degree.

a. Find three values of $\theta$, other than $81°$, such that $\cos \theta = \cos 81°$.

b. Find three values of $\theta$ such that $\cos \theta = -\cos 20°$.

c. Find three values of $\theta$ such that $\cos \theta = 0.3$.

d. Find three values of $\theta$ such that $\cos \theta = -0.48$.

# Find the Ferris Wheel

1. Imagine that the equations in parts a and b each describe the *x*-coordinate of a rider on some Ferris wheel in terms of time, where the rider is at the 3 o'clock position when $t = 0$. In these equations, $t$ is in seconds and $x$ is in feet.

    Recall that *angular speed* is the rate at which the Ferris wheel turns. In this situation, it is given in degrees per second.

    Give the radius, period, and angular speed of the Ferris wheel that each equation represents.

    a. $x = 25 \cos (10t)$

    b. $x = 100 \cos (3t)$

2. a. Write an equation that would give the *x*-coordinate of a rider on a Ferris wheel that has a smaller radius than the wheel in Question 1a but a greater angular speed.

    b. Describe how the graph for the equation in Question 2a would differ from the graph for Question 1a.

# A Trigonometric Conclusion

Congratulate yourself on a major achievement—finding out when the assistant should release the diver under the given conditions. Then consider this. Your work so far has involved a significant simplification of the problem. If the assistant uses the solution from *Moving Cart, Turning Ferris Wheel*, it could cost the diver his life. So there's still quite a bit of work to do on the Ferris wheel situation.

Before leaving the Ferris wheel problem, you will learn a bit more about trigonometry, including the use of polar coordinates and some important general principles called *identities*.

*Vick Chandra uses the Ferris wheel problem to understand more about trigonometry.*

# Some Polar Practice

*Polar coordinates* and *rectangular coordinates* give us two ways to describe points in the plane. This activity focuses on the relationships between the two systems.

1. a. Find the rectangular coordinates for the point whose polar coordinates are $(2, 30°)$.

   b. Find the rectangular coordinates for the point whose polar coordinates are $(5, 140°)$.

2. a. Find a pair of polar coordinates for the point whose rectangular coordinates are $(8, 2)$.

   b. Find a pair of polar coordinates for the point whose rectangular coordinates are $(4, -9)$.

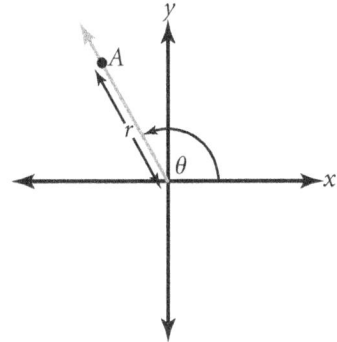

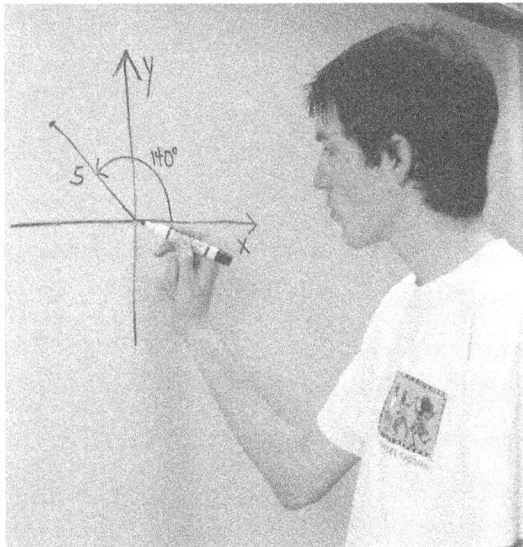

You know that the position of a point in the plane is usually described in terms of coordinates $x$ and $y$, which are its **rectangular coordinates** (or Cartesian coordinates). A point's position in the plane can also be described in terms of **polar coordinates,** usually represented by the letters $r$ and $\theta$.

For example, in the first diagram, point $P$ has rectangular coordinates $(4, 7)$. The variable $r$ represents the distance from $P$ to the origin. The variable $\theta$ represents the angle made between the positive direction of the $x$-axis and the ray from the origin through $P$, measured counterclockwise.

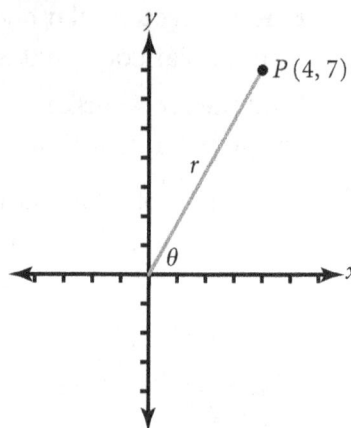

You can use the Pythagorean theorem to find that $r = \sqrt{4^2 + 7^2} = \sqrt{65} \approx 8.06$. To get $\theta$, you can use one of the trigonometric functions.

For example, you can use the sine function, whose general definition is $\sin \theta = \frac{y}{r}$. For point $P$, this equation becomes $\sin \theta = \frac{7}{\sqrt{65}}$, or about 0.868, which gives $\theta \approx 60°$. In other words, point $P$ can be represented approximately in polar coordinates as $(8.06, 60°)$.

The process can be reversed, starting from a point's polar coordinates and finding the point's rectangular coordinates. For instance, in the second diagram, point $Q$ has polar coordinates $(10, 240°)$.

You can use the general definitions of the sine and cosine functions to find the rectangular coordinates of point $Q$.

For example, $\sin \theta = \frac{y}{r}$, so $y = r \sin \theta$. Therefore, the $y$-coordinate of $Q$ is $10 \sin 240°$, or approximately $-8.7$.

Similarly, $\cos \theta = \frac{x}{r}$, which gives $x = r \cos \theta$. So, the $x$-coordinate of point $Q$ is $10 \cos 240°$, which equals $-5$. The rectangular coordinates of $Q$ are therefore approximately $(-5, -8.7)$.

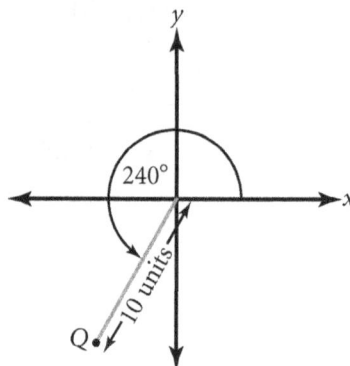

*continued* ▶

## Angles Greater Than 360°

The concept of polar coordinates is complicated by the fact that we don't restrict $\theta$ to angles between 0° and 360°. An angle of 360° or greater is simply interpreted as representing more than a complete rotation around the origin.

For instance, point $P$ can be represented in polar coordinates as (8.06, 420°), because a counterclockwise rotation of 420° from the positive x-axis results in the same ray from the origin as a rotation of 60°. Similarly, point $Q$ can be represented as (10, 600°), (10, 960°), (10, 1320°), and so on.

## Negative Angles

We also allow the polar coordinate $\theta$ to be negative, by interpreting a negative angle as a *clockwise* rotation from the positive x-axis. For example, point $Q$ can be described by the polar coordinates (10, −120°). The negative sign for the 120° angle indicates rotating 120° in the clockwise direction.

## Negative Values for $r$

The final complication for polar coordinates is that we allow negative values for $r$. If $r$ is negative, the point lies in the opposite direction from the point with the corresponding positive $r$-value.

For example, consider the diagram shown here. Suppose point $S$, in the first quadrant, has polar coordinates (2, 30°). Suppose point $T$, in the third quadrant, is in the opposite direction from the origin as point $S$ and is also 2 units from the origin. Point $T$ can be described by the polar coordinates (−2, 30°).

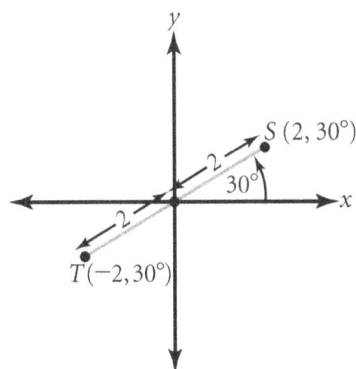

Note that point $T$ can also be described by the polar coordinates (2, 210°).

*continued* ⬧

## Summary: Multiple Representations

The use of arbitrary angles for $\theta$ and of both positive and negative values for $r$ means we can represent every point in the plane (except the origin) in polar coordinates in infinitely many ways. This creates some problems in working with polar coordinates, but it also leads to flexibility.

Because points have more than one representation, we sometimes speak of "a polar representation" of a point rather than "the polar coordinates" of the point. For convenience, you might refer to the representation with $r$ positive and $0° \leq \theta < 360°$ as the *standard polar representation.*

# Polar Coordinates on the Ferris Wheel

You may find it helpful to think of polar coordinates in terms of the Ferris wheel.

Picture a Ferris wheel with its center at the origin of the coordinate system. Then picture a rider on the circumference of the wheel, starting on the positive part of the x-axis and going counterclockwise.

In this model, the rider's r-coordinate gives the radius of the Ferris wheel. The rider's θ-coordinate gives the angle through which the rider has turned (starting from the 3 o'clock position). For example, a person on a 30-foot wheel who has gone one-fourth of the way around the wheel has polar coordinates (30, 90°).

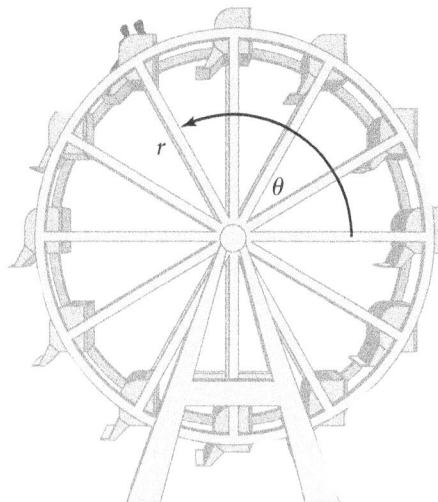

1. Suppose a Ferris wheel has a radius of 40 feet and a period of 20 seconds. The rider passes the 3 o'clock position at $t = 0$. Find the rectangular coordinates and the standard polar coordinates for the rider when $t = 3$, using the center of the wheel as the origin.

2. a. Find a value of $t$ different from 3 seconds, for which the rider would be at the same position as in Question 1.

   b. Use your answer to part a to find a different pair of polar coordinates for the position in Question 1.

3. Find general expressions for both the rectangular coordinates and the polar coordinates of a rider's position at time $t$. Use the Ferris wheel from Question 1, with radius 40 feet and period 20 seconds.

# Pythagorean Trigonometry

As you have seen, the definitions of the sine and cosine functions are based on a coordinate diagram like this one.

Specifically, to define $\sin \theta$ and $\cos \theta$, draw a ray from the origin that makes a counterclockwise angle $\theta$ with the positive $x$-axis. Then pick some point on that ray (other than the origin). Using $r$ to represent the distance from the point to the origin, $r = \sqrt{x^2 + y^2}$.

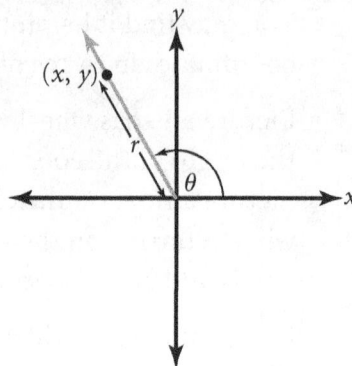

If the point has rectangular coordinates $(x, y)$, you define $\sin \theta$ as the ratio $\frac{y}{r}$ and define $\cos \theta$ as the ratio $\frac{x}{r}$.

Because the ratios $\frac{y}{r}$ and $\frac{x}{r}$ don't change no matter which point on the ray you choose, you can pick any point that is convenient. One common simplification is to pick the point that lies on the unit circle, which is the circle with radius 1 and center at the origin. Choosing this point simplifies matters because it means $r = 1$.

1. Suppose you choose the point $(x, y)$ so that it is on the unit circle. How can you express $x$ and $y$ in terms of $\sin \theta$ and $\cos \theta$?

2. What is the equation of the unit circle? That is, what condition must $x$ and $y$ satisfy if $(x, y)$ is 1 unit from the origin?

3. Use your answers to Questions 1 and 2 to write an equation relating $\sin \theta$ and $\cos \theta$ for points on the unit circle.

4. Choose four values of $\theta$, one in each quadrant. Verify in each case that your equation in Question 3 holds true.

# Coordinate Tangents

You've developed a way to define the sine and cosine functions for arbitrary angles. Now it's time to look at the tangent.

Remember that for a right triangle such as the one shown here, we define tan $\theta$ by the formula

$$\tan \theta = \frac{\text{opposite}}{\text{adjacent}}$$

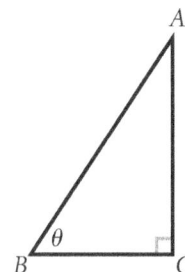

where *opposite* means the length of $\overline{AC}$ and *adjacent* means the length of $\overline{BC}$.

1. Suppose a point in the plane has rectangular coordinates $(x, y)$ and polar coordinates $(r, \theta)$, as in the second diagram. How would you define tan $\theta$ in terms of $x$ and $y$? Explain and justify your decision.

2. It's helpful to have equations connecting the various trigonometric functions. How can you express tan $\theta$ in terms of sin $\theta$ and cos $\theta$, rather than in terms of the coordinates $x$ and $y$? Think about how $x$ and $y$ might be expressed in terms of sine, cosine, and $r$.

3. Find each value, based on your definition in Question 1.
   a. tan 120°
   b. tan 230°
   c. tan (−50°)
   d. tan 385°

4. Sketch a graph of the equation $z = \tan t$, using $t$ for the horizontal axis and $z$ for the vertical axis. Include values for $t$ from −180° to 360° in your graph.

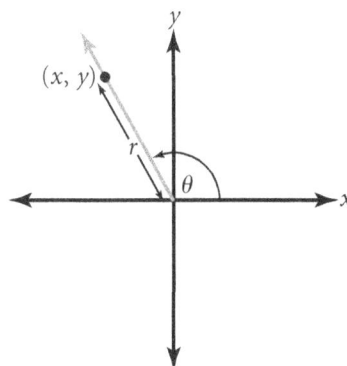

# Positions on the Ferris Wheel

In *Pythagorean Trigonometry,* you developed the equation $(\cos \theta)^2 + (\sin \theta)^2 = 1$. You saw that this equation is true no matter what value you substitute for the angle $\theta$.

Equations that are true no matter what values are substituted for the variables are called **identities.** In this activity and in *More Positions on the Ferris Wheel,* you will look at other identities involving the sine and cosine functions. The Ferris wheel model can help you develop and understand these identities.

The diagram shows two riders on a Ferris wheel, at the 1 o'clock and the 11 o'clock positions. The rider at the 1 o'clock position has turned 60° from the 3 o'clock position. The rider at the 11 o'clock position has turned 120°.

Suppose the Ferris wheel's radius is 50 feet. A rider's height, compared to the center of the wheel, is then given by the expression $50 \sin \theta$. But these two riders are at the same height, so $50 \sin 60° = 50 \sin 120°$. Dividing by 50 gives the following relationship.

$$\sin 60° = \sin 120°$$

The equation $\sin 60° = \sin 120°$ can be generalized, using the diagram shown here, to get an identity involving the sine function.

In this diagram, points $A$ and $B$ represent the positions of two riders on a Ferris wheel. That means $A$ and $B$ are the same distance from the origin.

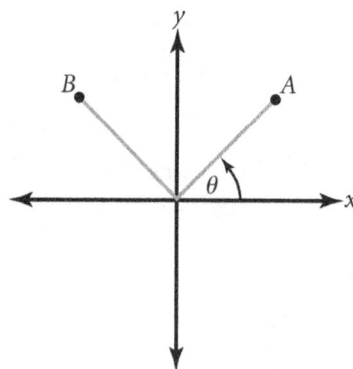

*continued* ▶

The angle $\theta$ represents the angle of turn for a rider at point $A$. Assume points $A$ and $B$ are at the same height on the Ferris wheel, so they have the same $y$-coordinate.

1. Find the angle through which the rider at point $B$ has turned. Express your answer in terms of $\theta$. First consider an example, such as $\theta = 20°$. Find the angle for point $B$ and then generalize.

2. Use the fact that points $A$ and $B$ are at the same height to write a generalization of the equation $\sin 60° = \sin 120°$.

# More Positions on the Ferris Wheel

In this activity, you will continue your exploration of trigonometric identities.

## Part I: Clockwise and Counterclockwise

The two Ferris wheel riders shown here both started from the 3 o'clock position. The first rider turned 30° and is now at the 2 o'clock position. The second rider turned −30° and is now at the 4 o'clock position. Negative angles, you'll recall, are interpreted as clockwise motion.

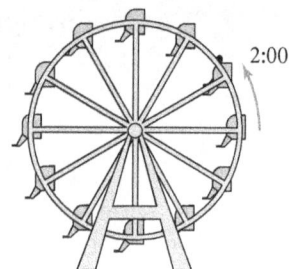

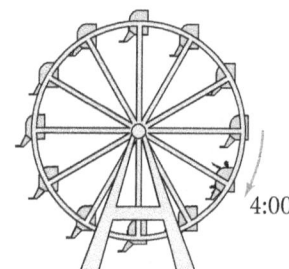

Assume the radius of the Ferris wheel is 50 feet. Recall that if a rider turns through an angle $\theta$, then his $x$-coordinate is given by the expression $50 \cos \theta$. Therefore, the first rider's $x$-coordinate is $50 \cos 30°$ and the second rider's $x$-coordinate is $50 \cos (-30°)$.

1. a. Explain why these two riders have the same $x$-coordinate. That is, why are they the same distance to the right of the center of the Ferris wheel?

    b. What does your answer to part a tell you about $\cos 30°$ and $\cos (-30°)$?

Now consider the general situation. In the next diagram, points $C$ and $D$ represent two positions on a Ferris wheel. In the case of point $C$, the rider has turned through an angle $\theta$. For point $D$, the rider has turned through an angle $-\theta$.

2. a. Explain why points $C$ and $D$ have the same $x$-coordinate.

    b. Use the diagram and your answer to part a to explain why $\cos \theta$ and $\cos (-\theta)$ must be equal.

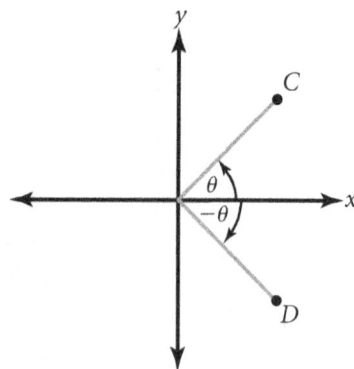

continued ▸

## Part II: From Identity to the Ferris Wheel

In Part I, you started with a situation on the Ferris wheel and generalized it to get a trigonometric identity. Now you will start with the identity and create a Ferris wheel explanation.

Consider the equation $\sin(-\theta) = -\sin\theta$.

3. Substitute values for $\theta$ to confirm that the equation is true for those values. Try a variety of values, including angles that are negative and angles that are greater than $360°$.

4. Create a Ferris wheel situation to explain the equation in a manner similar to the one you used in Part I.

5. Create a coordinate system diagram like that in Part I to illustrate this situation.

# A Trigonometric Reflection

In the activity *Moving Cart, Turning Ferris Wheel,* you figured out how many seconds after the Ferris wheel passes the 3 o'clock position that the assistant should let go of the diver to make sure the diver lands in the cart. Even though this is a simplified version of the problem, you couldn't have solved it without understanding mathematical concepts of trigonometry.

You first learned about trigonometry in the context of right triangles. Now you've found that the trigonometric functions can be defined for all angles. You have also learned some things about graphs, trigonometric identities, and polar coordinates.

Compile a summary of these ideas. Include diagrams as needed to help you explain any formulas. You don't need to include formulas that relate only to the unit problem, but do use the Ferris wheel to explain ideas about trigonometry.

# *High Dive* Portfolio

It is time to put together your portfolio for *High Dive*. Compiling your portfolio has three steps.

- Write a cover letter that summarizes the unit.
- Choose papers to include from your work in this unit.
- Discuss your personal mathematical growth in this unit.

## Cover Letter

Look back over *High Dive* and describe the central problem of the unit and the key mathematical ideas. Your description should give an overview of how the key ideas—such as extending the sine and cosine functions and finding falling-time functions—were developed and how they were used to solve the central problem.

In compiling your portfolio, you will select some activities that you think were important in developing the key ideas of this unit. Your cover letter should include an explanation of why you selected each item.

## Selecting Papers

Your portfolio for *High Dive* should contain

- *Moving Cart, Turning Ferris Wheel*
- A Problem of the Week

  Select one of the POWs you completed in this unit: *Tower of Hanoi* or *Paving Patterns.*
- *A Trigonometric Reflection*
- Other key activities

  Identify two concepts you think were important in this unit. For each concept, choose one or two activities that helped improve your understanding, and explain how the activities helped.

*continued* ▶

## Personal Growth

Your cover letter for *High Dive* should describe how the mathematical ideas were developed in the unit. In addition, write about your own personal development during the unit. You may want to address this question.

*How do you feel about your ability to solve a complex problem that has as many components as the "High Dive" problem?*

Include any thoughts about your experiences that you wish to share with a reader of your portfolio

# SUPPLEMENTAL ACTIVITIES

The supplemental activities in *High Dive* focus primarily on the trigonometric functions and their relationship with the Ferris wheel situation. Here are three examples.

- *A Shifted Ferris Wheel* examines how changing the starting time for the Ferris wheel would affect the function describing the platform's height.

- *A Change in Plans* presents an interesting twist on the original circus act problem in that you don't have to take into account the angular velocity of the dive.

- *Polar Equations* and *Circular Sine* continue the work with polar coordinates.

# Mr. Ferris and His Wheel

The Ferris wheel is named after its inventor, George Washington Gale Ferris.

Many of us have probably ridden on or watched a Ferris wheel at some point in our lives. But it's unlikely we know very much about its fascinating history. Here are some questions about the Ferris wheel that you may want to research.

- Who was George Ferris?
- Where was he raised and educated?
- How did he come to invent the Ferris wheel?
- What was the first Ferris wheel made of?
- How did the invention of the Ferris wheel change George Ferris's life?
- Did Ferris have any other notable inventions?

Write a report about your findings. You may want to expand your investigation to the broader topic of amusement park rides or to some other aspect of carnivals and fairs.

# A Shifted Ferris Wheel

In the main unit problem, the diver's platform is at the 3 o'clock position when the cart starts moving. Using this moment as $t = 0$, the platform's height after $t$ seconds is given by the expression $65 + 50 \sin (9t)$.

Suppose instead that at $t = 0$, the platform was at the 6 o'clock position.

1. Find an expression that gives the platform's height as a function of $t$.

2. Sketch the graph of the height function. Compare it to the graph for the main unit problem.

3. Consider other positions for the Ferris wheel at $t = 0$. Describe in general how changing the position affects the function describing the platform's height.

# Prisoner Revisited

Do you remember the prisoner from the Year 1 unit *The Pit and the Pendulum*? Well, he's back.

This time, he is lying on the table in the middle of a square prison cell that is 35 feet by 35 feet. A pendulum moves back and forth above him, as shown here. (There are no rats in this cell.) As before, a blade is attached to the end of the pendulum. The length of the pendulum does not change this time, but the table is gradually rising, moving the prisoner up toward the blade.

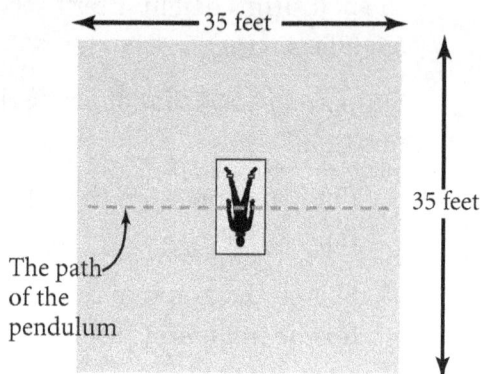

The prisoner notices that the pendulum's motion follows a sine-like pattern as it swings back and forth. Specifically, the pendulum's horizontal distance $p(t)$ from the center of the cell (measured in feet) is given by the function

$$p(t) = 15 \sin (60t)$$

where $t$ is the number of seconds the pendulum has been swinging.

The second diagram shows this horizontal distance, viewed from the front of the cell. (This diagram is not drawn to scale.)

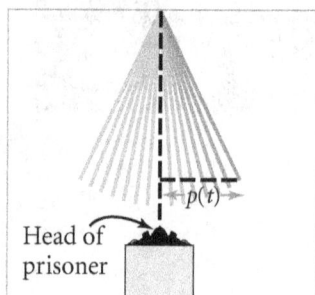

*continued* ▶

Suddenly, a friend of the prisoner appears in the adjacent cell. The friend rushes to the bars that separate the two cells. The pendulum swings alternately toward him and away from him, as shown in this overhead view.

The prisoner's friend realizes that if the pendulum comes within 3 feet of the bars between the cells, he can reach through the bars and grab onto the pendulum, stopping its motion.

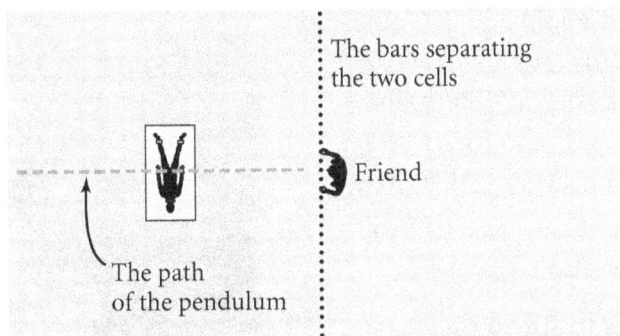

The bars separating the two cells

Friend

The path of the pendulum

1. Will the pendulum come close enough to the bars between the cells so that the prisoner's friend can reach it? Explain your answer.

2. If so, for how long will the pendulum be in the friend's range each time it passes by?

# Lightning at the Beach on Jupiter

As you may have noticed, the three variables *rate, time,* and *distance* are closely related.

In this activity, you are given information about two of these variables and asked to find the value for the third.

1. Light travels at about 186,000 miles per second. Jupiter is about 483,000,000 miles from the sun. (It's sometimes closer and sometimes farther away, but we'll use this average distance.)

   How long does it take for light to travel from the sun to Jupiter?

2. Amparo wants to spend the day at the beach, which is 100 miles away. She leaves at 8:00 in the morning and wants to be home by 7:00 that evening.

   For Amparo to have 6 hours at the beach, what should her average speed for the trip be?

3. You see a flash of lightning. About 6 seconds later, you hear the crash of thunder. Assume that the light reaches you instantly and that the sound travels at about 1100 feet per second. (The exact speed of sound depends on atmospheric conditions like temperature.)

   How far away was the lightning?

4. What general relationships exist among rate, distance, and time?

5. How do the concepts of rate, distance, and time relate to the main unit problem?

# The Derivative of Position

In *Free Fall,* you developed an important general principle about free-falling objects.

> If an object at rest falls freely from a height of $h$ feet, then its height after $t$ seconds is approximately $h - 16t^2$ feet.

This principle builds on the fact from physics that a free-falling object accelerates at approximately 32 feet per second for each second it falls.

Your task in this activity is to confirm the formula $h - 16t^2$ using derivatives. You will need to show that if an object's height fits this formula, then its acceleration must be 32 feet per second for each second it falls.

1. Suppose a certain object is moving downward so that its height $f(t)$ after $t$ seconds is given by the equation $f(t) = h - 16t^2$.

   a. At what rate is the object's height changing at $t = 1$? That is, what is the object's instantaneous velocity at $t = 1$?

   b. Explain why the number you found in part a is the same as the derivative of $f$ at $t = 1$.

2. a. Find the derivative of $f$ at $t = 2$, $t = 5$, and $t = 10$.

   b. Based on your answers to Questions 1 and 2a, give a general expression for $f'(t)$ in terms of $t$.

3. What does your result from Question 2b say about the object's acceleration?

# A Change in Plans

Preparations for the circus act are coming along nicely. You need only a little more time to work out all the details for the jump from a moving Ferris wheel. But the investors are getting impatient! They want you to start generating more income right now. So you decide to go on the road with a modified act.

Here is the plan for the modified act.

> You ask a randomly selected audience member to choose an angle between 0 and 360 degrees. You then turn the Ferris wheel that many degrees and stop it. When the wheel stops, the cart starts moving along the track toward the wheel from 240 feet away at a speed of 15 feet per second.

> At the appropriate number of seconds after the cart starts moving, the diver drops and lands in the moving cart!

To make this modified act successful, you will need a function whose input is any number of degrees between 0 and 360 and whose output is the number of seconds after the cart starts moving that the diver should wait to drop.

Find that function.

# Polar Equations

Over the years, you've worked with and graphed many equations involving $x$ and $y$. The graph of such an equation consists of all points whose rectangular coordinates fit the equation. For example, the point with rectangular coordinates (3, 4) is on the graph of $5x - 2y = 7$ because $5 \cdot 3 - 2 \cdot 4 = 7$.

You can also graph equations involving polar equations. As with equations using $x$ and $y$, the graph of an equation involving $r$ and $\theta$ consists of all points whose polar coordinates fit the equation.

For example, the point with polar coordinates (3, 90°) is on the graph of the equation $r + \sin \theta = 4$ because $3 + \sin 90° = 4$.

For each of these equations, first find some number pairs for $r$ and $\theta$ that fit the equation. Then use those number pairs as polar coordinates and plot the points they represent. Finally, use these points to sketch a graph of the equation. Find more solutions if you need them to get a good idea of what the graph looks like.

1. $r = \theta$

2. $r = \cos \theta$

3. $r = 2$

4. $\theta = 20°$

# Circular Sine

If you were to plot some points for the polar coordinate equation $r = \sin \theta$ and connect them, you might find that the graph looks something like the one shown here. It appears to be a circle, but it's hard to tell for sure simply by plotting points and connecting them.

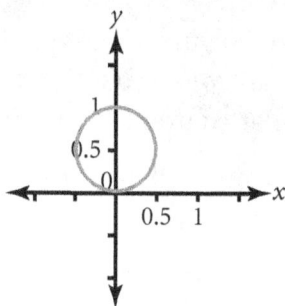

Your challenge in this activity is to show that the graph of the polar equation $r = \sin \theta$ is definitely a circle.

1. If this graph is a circle, what are the rectangular coordinates of its center and what is its radius?

2. What is the rectangular equation for the circle with the center and radius you found in Question 1? Suppose a point $(x, y)$ is on this circle. Use the Pythagorean theorem to get an equation for the distance from this point to the center of the circle.

3. How can you use the relationships between rectangular and polar coordinates to confirm that the rectangular equation for Question 2 is equivalent to the polar equation $r = \sin \theta$?

# A Polar Exploration

In this activity, you will investigate graphs and equations using polar coordinates and report on what you discover.

If you worked on the supplemental activity *Polar Equations,* you saw that the graphs of simple polar equations, such as $r = \theta$, can give very different graphs from simple equations with rectangular coordinates.

Here is one of the interesting graphs you can get from a fairly simple polar equation.

You might consult a trigonometry textbook or mathematics Web sites for ideas of interesting equations to explore. Your report should indicate any references you used and show clearly which ideas came from other sources and which are your own.

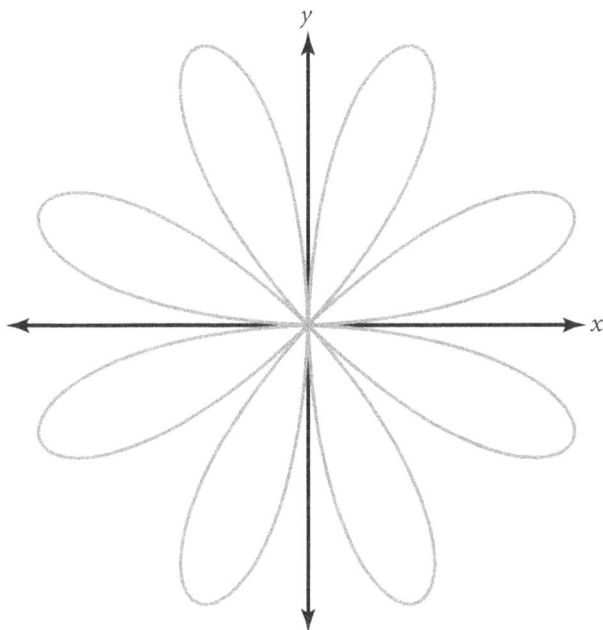

# A Shift in Sine

You have observed that the graphs of the functions $z = \sin t$ and $z = \cos t$ are quite similar. One way to describe the relationship is that if you shift the graph of the sine function 90° to the left, you get the graph of the cosine function. Alternatively, shifting the graph of the cosine function 90° to the right produces the graph of the sine function.

1. Express this relationship between the graphs as a trigonometric identity, writing $\sin \theta$ as the cosine of a different angle.

2. Prove the identity you found in Question 1. Use the relationship $\sin \theta = \cos (90° - \theta)$, which you verified for all angles $\theta$.

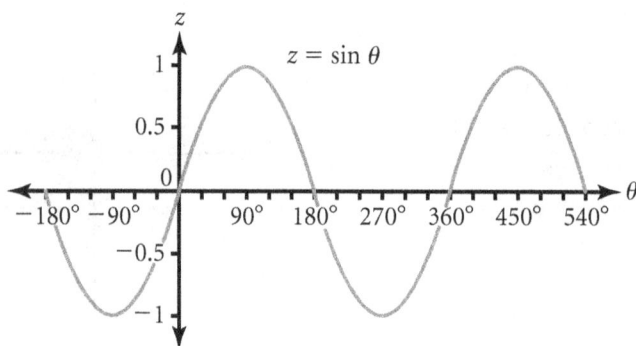

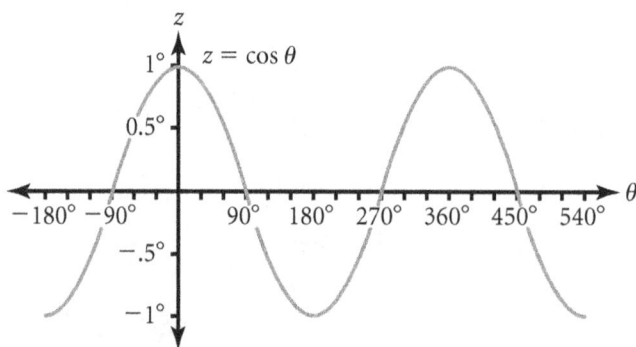

# More Pythagorean Trigonometry

The Pythagorean theorem tells us that in a right triangle such as the one shown here, the lengths of the sides satisfy the equation $a^2 + b^2 = c^2$.

The sine and cosine functions are defined for right triangles by the equations $\sin \theta = \frac{b}{c}$ and $\cos \theta = \frac{a}{c}$.

In the activity *Pythagorean Trigonometry*, you developed an identity involving the sine and cosine functions that resembles the statement of the Pythagorean theorem.

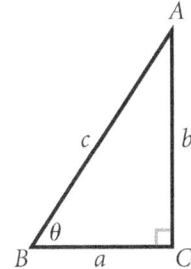

1. State that identity and explain it for acute angles based on the equations $a^2 + b^2 = c^2$, $\sin \theta = \frac{b}{c}$, and $\cos \theta = \frac{a}{c}$.

2. Develop similar identities involving the other trigonometric functions—tangent, cotangent, secant, and cosecant—based on the definitions $\tan \theta = \frac{b}{a}$, $\cot \theta = \frac{a}{b}$, $\sec \theta = \frac{c}{a}$, and $\csc \theta = \frac{c}{b}$.

# PHOTOGRAPHIC CREDITS

### Front Cover Photos

(From top left, clockwise) Jonathan Wong, Johnny Tran, Dylan Matthews, Thao Nguyen, Eden Ogbai

### Front Cover and Unit Opener Photography

Berkeley High School and Lincoln High School: Stephen Loewinsohn
Stock photos: iStockphoto

### High Dive

**3** Oxnard High School, Jerry Neidenbach; **5** FEV Create Inc/Getty Images; **12** Capuchino High School, Chicha Lynch, Hillary Turner, Richard Wheeler; **18** Shutterstock; **21** Shutterstock; **27** Lincoln High School, Stephen Loewinsohn; **31** Chigmaroff/Davison/SuperStock; **34** Harry How/Getty images; **35** Lincoln High School, Stephen Loewinsohn; **38** Photodisc/Getty Images; **40** Paul & Lindamarie Ambrose/Getty Images; **41** Shutterstock; **42** Shutterstock; **44** Capuchino High School, Peter Jonnard, Hillary Turner, Richard Wheeler; **48** iStockPhoto; **49** John Warden/Superstock; **50** Capuchino High School, Chicha Lynch, Hillary Turner, Richard Wheeler; **51** Hillary Turner; **66** Bettmann/Corbis; **67** Shutterstock; **70** Shutterstock; **71** Steve Vidler/SuperStock